Your Book of Traction Engines

The Your Book *Series*

- Acting
- Aeromodelling
- Animal Drawing
- Aquaria
- Astronomy
- Badminton
- Ballet
- Brasses
- Bridges
- Butterflies and Moths
- Cake Making and Decorating
- Camping
- The Way a Car Works
- Card Games
- Card Tricks
- Chemistry
- Chess
- Coin Collecting
- Computers
- Confirmation
- Contract Bridge
- Medieval and Tudor Costume
- Seventeenth and Eighteenth Century Costume
- Nineteenth Century Costume
- Cricket
- Dinghy Sailing
- The Earth
- Electronics
- Embroidery
- Engines and Turbines
- Fencing
- Film-Making
- Fishes
- Flower Arranging
- Flower Making
- Freshwater Life
- Golf
- Gymnastics
- Hovercraft
- The Human Body
- Judo
- Kites
- Knitted Toys
- Knitting and Crochet
- Knots
- Landscape Drawing
- Light
- Magic
- Maps and Map Reading
- Mental Magic
- Model Car Racing
- Modelling
- Money
- Music
- Paper Folding
- Parliament
- Party Games
- Patchwork
- Patience
- Pet Keeping
- Photographing Wild Life
- Photography
- Keeping Ponies
- Prehistoric Britain
- Puppetry
- The Recorder
- Roman Britain
- Rugger
- Sea Fishing
- The Seashore
- Self-Defence
- Sewing
- Shell Collecting
- Skating
- Soccer
- Sound
- Space Travel
- Squash
- Stamps
- Surnames
- Survival Swimming and Life Saving
- Swimming
- Swimming Games and Activities
- Table Tennis
- Table Tricks
- Tape Recording
- Television
- Tennis
- Traction Engines
- Trampolining
- Trees
- Veteran and Edwardian Cars
- Vintage Cars
- Watching Wild Life
- The Weather
- Woodwork

Your Book of Traction Engines

ALAN BLOOM

FABER AND FABER · LONDON

First published in 1975
by Faber and Faber Limited
3 Queen Square London WC1
Printed in Great Britain by
BAS Printers Limited, Wallop, Hampshire
All rights reserved

ISBN 0 571 10413 4

© *1975 Alan Bloom*

Illustrations

		page
1.	Mr Gerald Dixon of Sudbury, one of the first to 'rescue' a traction engine, inspects a 1909 Burrell which later began the author's collection	9
2.	A Ransome 'Portable' hauled by horses	13
3.	One of the earliest self-propelled traction or farm engines	33
4.	A threshing scene in 1911. Men forked sheaves of corn from the stack to the drum	37
5.	A more recent threshing scene but from the rear end. The protective cover for the man feeding the drum is an unusual feature. Straw is being baled	39
6.	This invention was known as the Walking Digger steam engine	42
7.	The driver's view of a Fowler ploughing engine	45
8.	Fowler ploughing engine: a side view of the gears driving the winch drum which carried up to $\frac{1}{2}$ mile of wire rope	45
9.	The complete view of a Fowler ploughing engine in operation, showing the wire cable being paid out	46
10.	Ploughs, mole drainers and cultivators: implements used by ploughing engines	47
11.	An Aveling road roller at work about 1912	48

12. An Aveling roller at a Rally	50
13. 'Ravee': a pioneer road locomotive built for work in India	54
14. A small general purpose traction engine built by Aveling and Porter	56
15. This Fowler, with its patent valve gear, was the forerunner of a larger and faster engine for road haulage	58
16. A representative of the lesser known makes of traction engine, this Allchin was built in 1911	59
17. A road haulage locomotive built by Robeys of Lincoln	59
18. A haulage engine with a rear-mounted revolving crane	60
19. These 12-foot driving wheels on a Fowler engine were a means of combating soft road conditions	60
20. An 1875 Robey hauls a new boiler to be installed in a Suffolk brewery	62
21. A Showman's Road Locomotive, with dynamo, built by Garretts of Leiston, Suffolk	64
22. An ornate and well cared for Burrell Showman hauling its train of fairground amusements	65
23. Two Foden timber tractors in a trial of strength at a Rally	68
24. A grand parade at a Traction Engine Rally	73

Introduction

Now that Traction Engines exist only as treasured relics of the past, the term has come to include almost every type of road-using vehicle. The dozens of Rallies held each year up and down the country use the term 'Traction Engines' on their advertising posters and programmes, but various machines which are not really engines appear there—steam waggons, rollers, ploughing engines and portables which are not self-propelled. Showman's or fair engines and general purpose agricultural engines, which are also seen at Rallies, can be more accurately described as traction engines, but their traction duties were limited. The 'general purpose' engines (more of which survived than of any other type of traction engine) were mainly used for hauling threshing equipment from farm to farm. In the same way, fair engines, with their gleaming brass and paintwork, hauled fairground equipment from place to place, but while the fairs were in progress they remained stationary, though they continued to work in order to provide the electric lighting. The remaining type of machine which you will see at the Rallies is the engine for road haulage, and this is the true traction engine, the forerunner of the motor lorry.

There were, as we shall see, a good many reasons why steam power was not used more for road haulage. It was not lack of inventiveness on the part of the pioneers that held it back, so much as frustrating laws, poor roads and competition from other means of transport. The first attempt to use steam on the roads was made over fifty years before the first practical railway locomotive appeared. During that interval and after, many weird and wonderful machines were tried out. Many did not survive, but nearly all made an impact of some kind, if only by inspiring awe or excitement in the minds of those who saw them.

Steam engines still inspire awe and excitement, even though the steam age has passed. With older people this may be due partly to nostalgia, but they have a fascination for many people of all ages—especially boys and men. I fancy this is due to something very basic in human nature. It is not only the sight of active power, of thrusting pistons and turning cranks and wheels, but the deeply evocative feelings aroused by a machine produced by man's inventiveness, in which two elements are harnessed to provide power. These two elements, of fire and water, had always played a vital part in the development of civilization. Both were good servants but bad masters, and each had its own important uses when harnessed. Fire and water in combination produce steam, but when James Watt and others discovered how to use steam as a source of power, it was a tremendous achievement indeed. Steam power was largely responsible for the Industrial Revolution, which led on to vast changes which have snowballed ever since.

Small wonder then that we look on these remaining examples of the age of steam with a sense of awe, excitement and, for that matter, pride, even if we are only onlookers. But the main reason why interest in steam engines has become so widespread and intense is their disappearance from everyday use. In my youth, all types—road, rail and stationary—were commonplace. I believed, as did most people, that steam power was here to stay,

1. Mr Gerald Dixon of Sudbury, one of the first to 'rescue' a traction engine, inspects a 1909 Burrell which later began the author's collection

and though I was always fascinated by the engines, it never occurred to me to study them intimately. When I bought my first traction engine in 1947, it was simply as a pet I was proud to own and drive. I had to justify its cost of £50 by working it on a saw bench and for land reclamation. Fifty pounds was then a large sum to pay for a traction, but mine—it was a Burrell—had been fully overhauled. At the same time hundreds of engines were being sold for as little as £10 and cut up for scrap.

By the 1960s, when over 90 per cent of the thousands of engines in existence before 1940 had been scrapped, interest in those remaining increased, and so did their value, as relics. I've heard many a moan from owners who had possessed traction engines before 1950 and sold them for less than £20. They would now be worth £2000. But to keep a sense of proportion we must remember that steam engines would not arouse so much interest or be so valuable today if the vast majority had not been scrapped. There would also be little demand for this book, which will include mention of some weird and impractical machines, as well as those that pioneered the development of the various types of traction engine of which examples have been preserved. These in their day played a vital role in road transport, constructional and agricultural work, for they provided a prime moving power far stronger, more reliable and more versatile than any that had been available before.

2

One of the principles discovered by the early inventors was that the power of steam varied with the pressure. James Watt realized the possibilities of steam, it is said, by observing how the lid of a kettle was raised after boiling point was reached. Such pressure was very slight indeed. He therefore designed a boiler so that when water boiled up into steam, it could only escape by pushing against a piston. When this was forced out, the connecting rod to a crankshaft transferred the power to a revolving wheel which then became the first source of power. James Watt, however, did not see beyond the stationary power unit, and although his engines could, with increased size of boiler, serve many purposes, he saw no point in aiming for higher pressured boilers. Perhaps he erred on the side of weight with safety as against the risks of explosion with high pressures, for in his day boilers capable of working even at fifty pounds per square inch were rare. James Watt was said to be so bigoted and bad-tempered that he discouraged anyone who tried to develop any other lines, and was against any idea of self-propelled engines.

But progress had to come and by 1800 several inventors had experimented with steam engines that could travel on the roads. Some of them were dabblers, but a Cornishman named Trevithick showed real inventive genius. His first steam carriage took seven or eight people, perched up high in a coach-like box, whilst underneath was a small but simple layout of boiler and cranks with geared sprockets driving the high rear wheels. In front a man sat above a single wheel, steering the vehicle with a tiller. A second road steam coach capable of carrying ten people, built by Trevithick in 1802, had hind wheels ten feet in diameter. But failure came, simply because capital was lacking for further development. No one was willing to back Trevithick and in disgust he turned from road- to rail-using engines.

It was Trevithick's genius that laid the foundation for future development, but none of the many attempts to produce road vehicles that could compete with horses over the next thirty years were really successful. It must be remembered that though improved means of road transport were needed, those were very difficult times. The Napoleonic wars, rioting and machine smashing, with much political unrest, were obstacles to investment in invention. Ten miles an hour was the fastest people could travel by stage coach, but at least they were fairly safe. Steam carriages appeared and services of a sort, mainly in London, were introduced—the forerunner of the buses and trams. But they were noisy, smelly and dangerous, and their frightening effect on horses can easily be imagined.

When the spread of railways began in the 1830s more settled political and economic conditions led naturally to exploring the possibilities of steam power for agriculture as well as for industry. The latter relied on heavy stationary engines, but farmers needed portable power, even if it was almost impossible at that time to consider steam as a substitute for the horse for haulage and field work. Horses could drag wooden carts, waggons and coaches over the rough roads of those days. They were also far more versatile for field work than any form of steam engine that could be devised. But would-be inventors

2. A Ransome 'Portable' hauled by horses

persisted, pursuing the idea that steam power had a great future, though sceptics thought it was only a dream.

    The impetus for further development came largely from farmers and landowners. They formed themselves into a society designed to improve, by whatever means they could, the output and profit from the land. This involved crops, livestock and machinery, and when the society became the Royal Agricultural Society in 1839, the decision was taken to hold annual Shows in different parts of the country. This was the kind of opportunity inventors needed—a centre at which new labour-saving devices could be exhibited not only for sale but to compete for cash prizes given by the Society for the best and most useful inventions.

The Ipswich firm of Ransome had established a reputation for farm implements, especially ploughs, and at the 1841 Royal Show held near Liverpool they exhibited their portable steam engine for agricultural use. This machine had a vertical boiler mounted on a wheeled platform with a flywheel to drive a small and rather crude threshing machine. A similar engine had a chain drive to a sprocket fixed to one hindwheel, which made it self-propelled, after a fashion. Steering it was another matter, for no better means were found than shafts in which a horse was harnessed, not to assist movement but to steer the front wheels. Even when the engine and boiler were improved as a power unit, steering remained something of a problem. For many years farm engines, even if self-propelled, were steered either by a horse in shaft, or by a man sitting in front turning a spoked wheel with a chain to the forecarriage below. The portable engine as a power unit gained far more popularity than these early attempts at self-propulsion. For little more than £100 farmers could make use of a portable for threshing either in the harvest field or in a barn, and if it took four horses to pull the unwieldy machine, which weighed as many or more tons, it could replace both horses and men when at work.

Every year there were improvements of one kind or another to be seen at work or on display at agricultural shows. As always, there was competition for both prizes and orders—and as always designers and inventors tried to improve on the work of others and sometimes succeeded. There was ample room for improvement, in making engines more efficient, more adaptable and, not least important, safer.

Amongst the difficulties these early inventors had to overcome, apart from reducing weight without loss of power, was how to keep steam from escaping to waste from the boiler, except at the piston-head power point. Glands that allowed the free movement in and out of the piston rod had to be so packed that steam did not come out as well. The valves which regulated the supply of steam to the pistons were equally important. This supply had to be strictly and accurately controlled by the operator, for a small fraction of an inch too much in opening a valve, even at such modest pressures as 50 lb. p.s.i., could treble the number of revolutions per minute. To some extent the invention of governors was a means of preventing an engine from shaking itself to pieces through allowing too much steam through the valves. These governors are the iron balls fixed to an upright stem above the boiler. As r.p.m. increased, centrifugal force made them open out as they revolved and in doing so they depressed the rod which in turn checked the valve parts from opening too widely. In this sense governors were an automatic regulator, though in practice they were needed only when an engine was working stationary. When it is travelling self-propelled, the driver alone controls the speed.

3

By the 1840s, years of experiment, trial and error were resulting in a more or less standard basic design. This, then, is the best place at which to explain how engines as we see them now emerged. To do this we should also see how fire and water became harnessed together to produce power most effectively. The fire needed to be placed underneath the boiler, so as to give the greatest possible heating surface. The first boilers, whether vertical or horizontal, contained a firebox with a door for stoking, and bars on which the fire was laid with an ash pan below, the flap of which could be used to regulate the draught to the fire. The chimney designed to go through the boiler and out the other end was a very early development as a means of providing extra heating surface to the water in the boiler.

It was only a matter of time before someone saw that to use several smaller tubes, rather than one larger chimney, would provide a greater heating surface. These tubes had to fit tightly into the tube plate at both ends, but at the opposite end to the fire a smoke-box was needed so that what came through could merge to escape through a single chimney.

The walls and roof of the firebox became heating surfaces only if they acted as an inner jacket, the space between it and the outer casing being filled with water. This important factor, of heating surface, naturally led to boilers being made long and narrow. Because of its advantages for portable and locomotive boilers, both for road and rail, the horizontal boiler eventually ousted the vertical type. Some, however, were a compromise between the two—a domed vertical boiler above the firebox and a horizontal one attached. Only the latter held fire tubes, which heated the water and made steam. But heated water and steam required room for expansion. Boilers of any type could never be full with water or there would be no room for expansion and steam.

The dome over the firebox evolved as a suitable container for steam under pressure and it was from this that the steam was conducted to the cylinders and through the valves to the pistons. Hornsbys of Grantham brought out one of these composite engines, which won first prize and £50 at the Royal Show held at York in 1848. The vertical firebox, domed above, needed no strengthening stays, as did a firebox within the outer boiler casing. These stays were like spacing bolts which prevented the inner and outer surfaces from buckling under heat and pressure, but as they needed to be steam- and water-tight as well as closely spaced they added considerably to the cost of manufacture.

Some early engines had the cylinders fitted inside the boiler, but they were better placed for working a crankshaft if they were fitted inside a steam dome or chest somewhere on top. Early engines had vertical cylinders, but an up and down movement of the pistons was also less adaptable for power that needed forward, horizontal motion. Sloping cylinders, fed by steam pipes from the boiler, were a half-way stage, until during the 1840s the horizontal was achieved. Some builders placed them beneath the boiler, and this kind of engine was known as undertype. For a self-propelled engine this would appear to be the obvious position, because the motion was between front and hind wheels, and gear sprockets to drive the latter were less complicated than on an overtype engine.

Yet the latter became standard for traction engines. One reason for this was the need for a large flywheel on which to place a driving belt for work. The flywheel also played an important part in smooth running. It was of heavy cast iron and fixed to the end of the crankshaft, which gave stability to the motion which was so easily excited by a trifle too much steam. We have already mentioned how tricky it was to manipulate the regulator handle so that the cylinders received enough but not too much steam, and the flywheel absorbed or cushioned some of the thrust. The flywheel also very often became the means of setting the cranks in motion. Sometimes the position of the piston within the cylinder was such that, when steam was allowed in by using the regulator rod, the piston would go neither way, because space on one side or the other was too confined. But a pull on the flywheel by the driver provided this space and steam then began its work.

This is the difference between the pistons and cylinder of a steam engine and those of a car. Force is exerted on both sides of the former in turn, but on one end only of the latter. The steam cylinder is therefore a compensating one. As it moves one way one valve is closed and another is opened. Its momentum, fixed to a crankshaft, ensures its return, and so again one valve port is closed and another opened, as pressure from steam pushes against both faces of the piston in turn. It is shaped like a thick round plate (or disc) on the end of a rod which sends the thrust on to the crankshaft.

This crankshaft, to which the big end bearing of the piston rod is fixed, is also used for the valve mechanism. The different types of valve gear need not be discussed here. All that need be said is that they could never be other than automatic. All are geared in some way so that steam is admitted to the cylinders precisely at the point when it is needed to push the piston the opposite way from which it has come. The distance it travels from one end to the other of the cylinder is termed the stroke. Naturally, the larger the engine the longer the stroke, and the greater the diameter of the cylinder with its close-fitting piston within, the more steam is required.

These are elementary factors governing compensating steam engines of all types. Other vital factors which had to be invented and perfected included the guiding of the piston rod. It was necessary to keep this working in line with the cylinder, and to allow for the rise and fall of the crankshaft there had to be a connecting rod. One end—the 'big end'—rose and fell as the crankshaft revolved, but the 'little end', where it joined the piston rod, did not. As a bearing, movement of the little end was only slight, but it had to be free. More than this, it had to be free within a limited space—no more and no less than the length of the stroke. To ensure this a strongly fixed frame, known as the cross head, played a vital part, as did the slide bar on which the little end was carried.

All these moving parts depended on oil for smooth running. Where a constant supply was needed, drip feed cups were fitted, but for the cylinders, where metal was really hot from live, dry steam, a much heavier oil was required. This supply was so vital that force-feed lubricators, worked by an attachment to some moving part, became standard equipment.

From the very beginning, some means had to be found of supplying water to an enclosed boiler, under pressure. Again this had to be automatic, worked by the motion of the engine, but capable of being switched on and off. The early pioneers installed a pump which with a short stroke drew water from a tank and forced it in against the pressure in the boiler. Its entry was through a 'clack box' which, in other words, is a non-return valve.

It was just as important to know when the boiler needed replenishing as when to cut off the pump, so the gauge glass was introduced. It was fixed where the driver could keep it under his eye and it consisted of a vertical glass tube held by brass fittings with taps, so that the water level in the boiler corresponded with the level seen in the glass.

Two safety devices were also standard equipment by the 1840s. We can imagine the kind of accidents which made them vitally necessary, knowing how terrifying boiler explosions must have been. The first device was seen to be necessary very early on. It was some means of allowing steam to escape if pressure built up beyond the strength of the boiler plating. Fitted to the dome, or wherever steam built up to its maximum, the valve lifted when a spring set at a given tension gave in to the pressure from below. It could be adjusted by increasing or releasing the tension, either by turning a screw or by moving a weight hung on an arm.

The second device was a fusible plug, fitted to the roof of a firebox which was riveted and stayed inside a boiler. The effectiveness of the firebox depended on being able to transmit heat to water behind the casing metal, but if the water level dropped so that there was none above the fire, on what is called the 'crown', then heat would buckle or split the metal. Explosions of this kind were due to negligence on the part of the driver and, as a safeguard, a fusible plug had to be fitted. This was a screwed plug, of sufficient size for a small hole to be bored through the centre. This hole was filled with lead and, because lead melts at a low temperature, an absence of water on the crown led to melting lead and a jet of steam directed on to the fire. To 'drop the plug' ranks as a disgrace to engine drivers even today, because it need never happen. But errors of judgement do occur and though an explosion is far worse, the nuisance of having to take out the fire, wait for the metal to cool, remove the firebars, and unscrew and re-lead the fusible plug before re-filling the boiler and relighting the fire, are deterrents to negligence. I can say this having experienced the shame as well as the tediousness of the task, which took four hours before my engine was mobile once more.

    The regulator and the reversing lever are the two principal points to be used by a driver. The first regulates the amount of steam into the cylinders; the second controls the point of entry. It is simply a matter of reversing the valve settings and consists of a rod fixed to a lever on a notched quadrant. These notches make it possible for the length of the stroke to be varied, in both forward and reverse. When the lever is notched to the centre of the quadrant, it cannot move either way, for no steam can enter. This is the neutral or stop position.

4

By 1850, several firms were well established as makers of traction engines. These were still mostly of the portable type, because self-propulsion was still far from perfect, apart from the additional cost entailed. These firms had seldom begun as engine builders. It was more often a case of a business associated with engineering in some way branching out to meet a growing demand. Some were iron foundries; others were specialists in farm implements and a few businesses even grew up from such humble beginnings as that of blacksmith.

Most builders who launched out into steam engines were, however, situated in agricultural districts. A good ninety per cent were in the eastern half of the country, partly because there the need for modern farm equipment was greatest. In much the same way, builders of stationary steam engines were established in industrial areas. But all these firms made engines with what seems a very low pressure, and it would be only guessing if we said this was what they believed to be quite adequate. No doubt they had the safety factor well in mind, but even Ransomes' 'Farmers' Engine', a much praised introduction to the Leeds 'Royal Show' in 1849, was built to run at 45 p.s.i.

This engine was self-propelled. What was more, it was steered from the footplate by rack and pinion. Its two cylinders were placed underneath the smoke-box, just clear of the steering gear of the front axle, with forked connecting rods turning a crankshaft just forward of the hind wheels. Not only was a form of clutch fitted so that it could be run for belt work when not in road gear, but it had two speeds as well. This was achieved by having a counter shaft on which toothed gears or sprockets gave different ratios. It is the principle on which even modern cars work—of altering the speed of the vehicle according to which sprocket takes the actual drive. By comparison, steam engine gears were clumsy, as these sprockets could not be engaged whilst revolving, or disengaged whilst under power.

The 'Farmers' Engine' carried fuel and water in tow, not as tank or bunker forming part of the machine. The footplate on which the driver stood was just a platform. But it had springs above the hind wheel, which were scarcely ever used for agricultural engines later on. Apart from the basic difference of having the cylinders and motion underneath the boiler, the steam dome was well forward on top of the 5′ 6″ long boiler barrel. The makers had broken away from the domed firebox and had simply rounded the top, level with and towards the boiler.

Most of these features were quite novel. What was more, the engine not only won first prize that year, but came through a series of tests with flying colours. It was given several days' work threshing corn. This was hard work and yet it used just under $\frac{1}{2}$ cwt of coke and forty gallons of water per hour. It could raise steam to working pressure of 45 p.s.i. from cold in forty-five minutes, take fairly stiff hills at walking pace in low gear, and reach ten to twelve miles an hour on the level. Ransomes claim that its lightness (working weight $2\frac{1}{2}$ tons) and its versatility received very favourable comments in the Press. And yet, in spite of all this, their 'Farmers' Engine', on which they placed such great hopes, proved to be a failure. Within two years production came to a halt because orders for it had almost ceased. They must have wondered why, but the true reason did not emerge for many years. It was mainly because of farmers' prejudice and conservatism. The engine was in fact too far ahead of its time. Farmers were suspicious of self-propulsion and steering from the rear. But it was not the fault of the engine itself that it proved too light for the dreadfully poor roads of that period. When self-propulsion was tried again many years later, roads were still bad enough to need engines with much broader tyres and other devices to combat mud and soft earth tracks. This meant much greater weight, and a much greater spread of weight on the surfaces over which they had to travel.

It is surprising that the Ransome design was not made the basis of the design which became standard at that time. The latter had cylinders and motion placed on top of the boiler, and it may well be that the undertype was discarded because of the danger of sinking in pot-holes and soft earth. It is not difficult to imagine the trouble it would cause if the crankshaft dipped into earth or mud because the engine wheels failed to keep on hard level ground.

The lack of enthusiasm for self-propelled engines in general and underslung motion in particular did not, however, dampen the widespread faith in steam power. The Royal Agricultural Society was still offering incentives, and builders of engines up and down the country were often completely committed to competing for prizes. These were not only for design and efficiency, but included economy as well. For example, tests were arranged annually so that comparisons of coal and water consumption could be made. A formula was drawn up on the basis of the coal consumption per hour, per single horse power, of the various engines competing. The prize-winning Garrett Portable in 1848 burned $11\frac{1}{2}$ lbs of coal per h.p. per hour. The same Leiston (Suffolk) firm again won in 1850, having reduced the figure to $7\frac{1}{2}$ lbs by studying wasted steam, but the next year, Hornsbys of Grantham had cut it still further to $6\frac{3}{4}$ lbs and again in 1852 to $5\frac{5}{8}$ lbs, and Claytons of Gainsborough topped the economy table in 1853 with $4\frac{1}{3}$ lbs. This was an important factor, even if coal cost only a few shillings a ton. We must never forget that £1 at that time was roughly equivalent, in purchasing power, to £15–£20 now.

At the Great Exhibition in Hyde Park in 1851, there was a machine called a mole drainer which could only be operated by steam. It was shown by John Fowler, a young man who had great vision and a very inventive mind. He had already used his mole drainer with great success. It was simply a low frame with four wheels between which a blade cut deeply into the soil, drawing a bullet-shaped plug so that a space rather like a mole's tunnel remained into which excessive soil wetness could run. It worked best in clay soils, which were of course most in need of draining.

The engines used by John Fowler for his drainage schemes were Ransome portables. But he had ideas about how to use steam for ploughing, as well as other aspects of farming. In 1856, still in conjunction with Ransomes, who were the leading plough-makers in Britain, his first ploughing outfit was put on trial with great success. The principle was of a powerful portable engine at one side of a field with a wire cable attached to a portable winch on the far side. Between the two, a double or two-way four-furrowed plough was drawn back and forth, and a gearing system on the anchor winch moved it as the work progressed. As with the mole drainer, a steersman was needed on the plough; on reaching one end of the area being ploughed, he changed his position so that the plough would reverse with a see-saw motion, without having to turn. This first design of a steam-drawn plough remained roughly unchanged for the rest of the century. The methods of drawing it were the subject of many a trial and failure, but in due course the steam-ploughing sets which were in widespread use practically all bore the name-plate of John Fowler of Leeds as the maker. The firm he founded became supreme in this field.

Clearly the problems of soft earth and bad roads must have been a constant worry affecting both builders and users of engines for field work and travel. Although the width of tyre was an important factor, weight was equally vital and one often went with the other. Light weight meant low power, less stability, and more power and more stability meant greater weight—a rather vicious circle.

And then, in the 1850s a man named Boydell came up with a startling invention—a wheel which was given his name. James Boydell first applied for a patent in 1846, but it was not until eight years later that he perfected his wheel sufficiently to persuade engine builders to take notice. By 1857 Garretts, Burrells, Tuxfords and Claytons were turning out engines fitted with the Boydell patent wheel. The general idea was that as the engine moved, the wooden flaps fixed to the wheels fell into position on the ground so that the weight was spread. It was, in fact, a form of tracklaying process, not continuous like modern tracklaying or crawler tractors, but with each wheel carrying its share of the weight independently. These Boydell-wheeled engines were of course self-propelled, with chain drive. They were mostly built up as a safer version of the more or less standard portable design; because the wheels did not sink or cut deeply into soft surfaces, the pulling power of the engine itself was far greater. In some cases only the rear wheels were fitted with the flaps, in others all four, but Tuxfords' model of 1858 was a three-wheeler of rather outlandish design, weighing twelve tons.

The prices of such engines were steep, compared with the portable of the day. They varied from £550 to over £1000 for the Tuxfords, according to size and horse power and whether there was a single driving wheel or two. Burrells charged £750 for theirs and it was this famous firm at Thetford which went all out to perfect and popularise the Boydell Patent. Many were exported to the West Indies, South America and Russia, and even the British Army began using them for hauling heavy field guns. On one trial a Burrell engine hauled a forty-three-ton load up hill and down again, to the astonishment of all who witnessed it.

For a few short years, it was believed that the old problems of weight, power and traction for road and field work had at last been solved. It seemed to make sense—for an engine to provide automatically its own firm level track on which to run. Yet by 1860 Burrells were the only remaining firm which had not lost faith in the Boydell patent wheel. The last one was built in 1862 and in June that year Boydell died in the bitter knowledge that his invention was about to die as well. He must have known that another inventor named William Bray of Folkstone had also been working on the same problem, at about the same time. Bray, however, put less emphasis on weight-spreading and more on giving driving wheels the means of traction or grip. No doubt his belief was that provided an engine could obtain a grip, whether on muddy road, track or soft earth field, it could keep out of trouble and utilise its power fully.

William Bray, an erstwhile marine engineer, therefore invented a broad-tyred hind wheel, not fixed to the axle but rather like that of a free wheel on a bicycle. There was more to it than this, of course, but the problems of grip were solved by having teeth fitted in the wheel surface. These were cleverly connected so as to move in and out as the wheel revolved, thus biting into the ground at the correct point. The ratio of weight to power was much more favourable than that of the Boydell principle. The engine weighed six tons against Boydell's ten tons. It could carry 8 cwt of fuel, which was enough for a day's work, as well as two hundred gallons of water. When its power was demonstrated, with a driver behind a steersman in front, people were amazed at the tonnage it would pull, at 60 lbs p.s.i., which was the working pressure.

The Bray engine was featured in the *Illustrated London News* in 1858, and even the British Admiralty ordered one for trial. In 1859 an American circus used an ornamented Bray engine to haul a train of vehicles to its site with a great flourish, to become in a sense the first 'Showman's' road locomotive. Other Bray engines—some built on outside contract—were used to haul heavy loads to and from the docks, and the prospects of real success seemed very bright. An extra large one was shown at the Great Exhibition in 1862; with twin cylinders, weighing twelve tons, but with pressure of 90 lbs p.s.i., it could haul forty-five tons with ease. Another was multi-purpose, for it included a winch and a crane, and yet another was built for export for Turkey. This very much resembled the then design of a railway locomotive, except for the wheels. And yet, for all the fanfare and approval, the Bray engine failed in the end. Its failure was gradual rather than sudden as in Boydell's unhappy case. The latter's engine did no damage to road surfaces, but the flaps soon wore out. The greatest snag about Bray's, one can imagine, was the damage caused to the roads by the iron teeth projecting from the wheels.

One more of the weird and wonderful machines of this period must be mentioned. This was the 'Steam Elephant' which James Taylor of Birkenhead first produced in 1859. It was quaint but very compact, with a funnel chimney out of all proportion to the short boiler barrel and firebox. So were the driving wheels with elaborate mud splashers, and driving, stoking and steering were all from the chimney end. In spite of its size of only four feet four inches high, engine and boiler only four feet eight inches long and two feet nine inches wide, and with a working pressure of fifty-five p.s.i., its two small cylinders provided enough power to haul eleven tons. It was also claimed to be capable of turning on a sixpence. More serious claims by the builder, as well as expert approval, should have resulted in this design becoming widely popular. It was produced with modifications for three or four years, and then virtually disappeared. At least, no further records of its being used have survived, and we can but assume that it too was found to be lacking in certain basic essentials.

One can be pretty sure that several other inventors and builders of this period fell by the wayside. Some had more enthusiasm than experience; their engines proved impractical and they never made the headlines. Others, like James Taylor, must have been forced to give up in bitter disappointment. Those remaining were mostly firms well established as engineers, who could more easily switch to the production of other and more profitable machines. The contest between all these lasted thirty years or so until, finally, about a score of firms were producing and selling engines of proven reliability.

5

The final failure of both the Boydell and the Bray systems made it seem that after all steam for road haulage was not practical. By that time, farmers had begun to realize that they had a challenge to meet. It was the challenge of cheap imported grain and the competition for workers with industry, which could pay higher wages than they could pay their own labourers. Hence the need to reduce costs of production, which could be achieved only by greater mechanization. For over a century this has been a constant battle for farmers, but in the 1860s steam was still the only source of power other than men and horses.

This need for more power on the land, especially for harvesting, was the incentive for builders of engines. One young man, named Thomas Aveling, began farming on his own account on Romney Marsh in Kent, which was rather like the Fen country his forbears had come from. He had a mechanical turn of mind and set about improving any implement he saw that was capable of being improved. He also recognized the potential in steam. Portable engines were being turned out by a dozen or more builders to meet the demand, but they were still pulled by horses. Tom Aveling decided that if only horses were not needed to haul these engines from place to place, there would be an enormous saving.

In 1859 he took out a patent for a self-propelled engine of his own design. It was an adaptation of a Clayton and Shuttleworth portable he had bought, and though it was not the perfect answer, the idea was good enough to persuade Claytons to build one to his specifications. It was chain-driven to a sprocket on a rear wheel, but this was rather long because the cylinders were above the firebox and the crankshaft close to the chimney. His steerage was also patented. It consisted of a single wheel on a projecting forked frame on which the steersman sat with legs dangling as he worked the tiller. This avoided the need for a horse in shafts for steering, on the principle that the single wheel acted as pilot for the pair placed under the smoke-box. Where the one went the others had to follow.

This steerage worked well enough to be used when in 1861 Aveling decided to build his own engines at a works he set up in Rochester. The first new engine he turned out was a reversal of the power thrust previously used, which was more or less standard on portable engines. By placing the cylinders forward and the crankshaft behind, closer to the rear wheels, the chain drive was shortened and simplified. The 1861 engine had its cylinders incorporated in the smoke-box, which prevented loss of heat. The need for lagging or insulation had long been met wherever possible. It stood to reason that contact with cool outer air reduced heating capacity within. Boilers were usually lagged with close-fitting narrow strips of wood lengthways with the barrel, with an outer wrapping of sheet metal—a practice that continued more or less to the end of the steam era.

Thomas Aveling took out patent after patent. This was necessary because, at a time when new improvements were badly needed, he had to take every means of preventing others from copying and cashing in on what must have often been costly experiments. Under a patent, the law could be invoked against anyone who copied a design, but this was not altogether effective. Ways of making slight modifications to some patent part could be found and these could well lead to further improvements. It was all part of the march of progress. Patents were needed just as much for certain components of engines, no matter what purpose they served, and this included the link chain which Aveling used as a drive. It was on the same lines as a bicycle chain, and no doubt even this had its origin in the brain of some engine man of long before.

Another feature of Aveling's new self-propelled engines was their capacity to carry both coal and water. The tank, the top of which was the driver's footplate, and the coal bunkers above and behind, were in fact the prototype for what became standard. Within twenty years every engine builder was using this space in the same way.

Amongst the difficulties overcome during the 1860s was that of wheel construction. Builders had begun with wood, iron shod where necessary. Then they went in for cast iron, which was less prone to wear. But it was more likely to fracture, so that gradually wrought iron, sometimes using angle iron or T. iron for strength, was incorporated. Finally the use of a combination of cast iron hubs on forged axles, with wrought iron spokes embedded in the casting, became a widespread if not standard practice. Heavy castings on which the spokes were riveted formed the basis of the tyres. These were of plate iron, bent to the correct circle to make it flush. To finish and to hold all this outer metal together, as well as to assist the grip, strakes were used. These consisted of bars, in which countersunk rivets were inserted so that they finished flush. These bars or strakes were finally found to be most effective if fixed obliquely to the tyre and not at a right angle. Riveting was of course standard practice wherever plate metal had to be held together. There was no welding in those days, except by what could be done in a forge. Plate work

3. One of the earliest self-propelled traction or farm engines

could not be, and the metal had therefore to be overlapped and riveted. This applied to boilers and fireboxes, smoke-boxes, watertanks and coal bunkers.

Boilers were of course tested as a matter of routine before being fitted. All openings, such as 'mud hole doors' for cleansing, had to be tightly closed and the boiler filled with water. A special hand pump was then attached which forced even more water in through a non-return valve. This increased the pressure, which was registered on a gauge, and if the boiler maintained a pressure for an hour or more of about fifty per cent greater than that at which it was designed to work in steam, then it would be passed. The question old engine men ask even now is not at what pressure does the safety valve blow off, but 'what is she pressed at?', meaning the working pressure for which the engine was designed; they know that the margin of safety is well beyond this figure.

Chain-drive traction engines were the most usual type during the 1860s. One other difficulty overcome was how to arrange a drive on to the axle so that in turning, one rear wheel had to travel farther than the other, if the axle was fixed. One of the earliest inventors of road engines in 1826 had found what proved to be the real answer. It was in the bevel pinion differential gear. Various other methods had been tried, but this was the answer, and it has been used ever since, for even the stub axles of a modern car are housed in an axle box with bevel pinion gears to provide the necessary differential for turning and cornering.

On traction engines, the chain-drive from the crankshaft to the driving wheel sprockets had its faults. It was liable to become slack and adjustment devices were imperfect. In other words, it suffered from wear and tear, and a more reliable alternative had to be found. Gearing was the only alternative. But the position of the crankshaft, even when this was at its nearest point to the hind driving wheel, was still too remote for a direct connection to be made. This applied as much with the now rarely used undertype engine as with the more usual arrangement of front cylinders and rear crankshaft above the boiler. To fill the gap, a second shaft was fitted, so that in turning the crankshaft pinion bit into the cogs of a lower placed shaft, and the gear wheel on this transferred the drive to the driving wheel sprocket.

This idea was capable of expansion. Second shafts came to be fitted, all contributing to the required gear ratio and the need for fully controlled drive. It also became the means of providing gears for controlling speeds, so that low and high gear could be used according to need. If the engine was moving dead slow in some yard, bottom gear gave greater steering control as well as greater pulling power, but a top gear made longer journeys much quicker. This may seem elementary, but a century ago it was very much a modern development. The shafts and gears were cast iron, and all carefully worked out so as to achieve the correct ratios for the speeds required. Engines came to be described as two, three or four shaft, terms which only the initiated understood.

The time came when three forward speeds were installed for road locomotives and showman's engines, but changing gear was a very different procedure from that in a car with a clutch. A steam engine had no separate clutch. The drive was direct, and if the crankshaft pinion was slotted into a driving shaft, the engine had to be in motion. But unlike the railway locomotive, on which the power is permanently connected to the driving axle from the piston rods, the traction engine could be taken out of gear. This could be done only with the steam cut off and the motion still, or nearly so. A lever, pushed aside, drew mesh gear away from that of the crankshaft. This left the motion out of gear and it was in this position that farm engines spent most of their time at work. Out of gear, with no drive connection to the wheels, the engine worked with a belt on the flywheel, powering a threshing set or sawbench, in just the same way as the simpler, cheaper portable steam engine.

The time also came when some indication of the power of an engine was necessary. The size of the cylinder, in conjunction with the steam-raising capacity of the boiler, determined the power. As a basic comparison 'horse power' was used as a measuring unit, just as candles were used for measuring artificial light. But steam engines could not be rated on pulling power, because so much depended on road wheels obtaining a grip.

The greatest power existed and was often used on the fly wheel or when the engine was working stationary. A true comparison with horse power (which was tractive) could not therefore be made, so the word 'Nominal' was used as a prefix whenever the actual power had to be quoted, and it was based on the size or cubic capacity of the cylinder. This was of cast iron, and though the bore for the piston would determine the power the h.p. rating was in fact nominal. Ratings varied from 4 to 8 n.h.p. for small haulage and general purpose engines including 'portables' and went up to 18 for ploughing engines. These figures would be multiplied several times if a true comparison with horses was possible.

Gearing has always been a case of one sprocketed or toothed wheel cogging in with another of different size. A large one cogging on to a smaller one increases speed by ratio, just as a small one on to a large one reduces it. When driving with a belt on the flywheel to a threshing drum, the revolutions on the drive shaft of the drum could be much faster than those of the flywheel of the engine. The latter was three or four feet in diameter, and that of the thresher one foot or less. It was the engine driver's job to see that all ran smoothly when a threshing set was at work. By the 1860s, threshing contractors were catering for those farmers who could not justify the cost of a machine of their own. Many would have no more than a few days' work for it, in the whole year.

The introduction of the threshing machines was revolutionary, for in one day the crop from about thirty acres of corn could be threshed out, depending of course on the yield. But a dozen or more men were needed when the 'threshing tackle' arrived. The driver had to keep his engine stoked, watered and oiled, and was also in charge of the threshing drum, elevator or chaff cutter. He was the key man, and any failing of the machinery

4. A threshing scene in 1911. Men forked sheaves of corn from the stack to the drum

would make all the other men idle. The bond cutter or feeder cast the sheaves into the mouth of the drum, and he was dependent on the men forking up sheaves of corn from the stack or carts. Then there were those who stacked the threshed out straw which came from the drum to fall into the elevator. This was often called the 'straw jack', and would be altered in pitch so that as the straw stack grew it would still carry on, supplying the stack builders. Two or three men were needed to dispose of the grain. This trickled out from the end of the drum nearest the engine, and they had to beware of the endless belt speeding between the two. There were usually four trap doors for the grain, which made sure that while the engine was working, there were always two corn sacks attached by hooks so that none could spill. When a sack was full, the trap door had to be closed, but not until the one next to it, on which an empty sack was in position, was opened. The other two trap doors were for the 'dross' or 'tailings'—small weed seeds and low grade corn. This grading was made possible by the use of sieves or riddlers inside the drum.

Grain sacks were of a standard size, and the man in charge of filling them from the drum knew what each filled sack should weigh. Wheat grain was heaviest at 18 stones, whereas barley was 16 stones and oats the lightest at 12 stones per sack. The task included weighing to the correct amount and keeping a tally, before sacks were tied up and taken away.

By far the worst of the many tasks involved in threshing was that of the 'chaff boy'. His duty was to see that the husks and tiny pieces of straw blown down the chute of the machine were bagged up and kept clear. Mostly he had to work in the narrow space between the corn stack and the threshing drum—a wind tunnel for dust and dirt, close beside half a dozen moving wheels and belts. Chaff bags were light, but he had to carry

5. A more recent threshing scene but from the rear end. The protective cover for the man feeding the drum is an unusual feature. The straw is being baled.

them, often two at a time, to a building where they were tipped out, and then hurry back for more. In those days, men and boys alike were expected to work up to twelve hours a day except during the darkest winter period.

The system of contracts for threshing was a natural outcome of the development of steam on the farm. A complete outfit cost close on £1000—equal to nearly £10,000 today—so very few farmers could afford to buy one. Many builders of engines—Burrells, Ransomes, Claytons, Marshalls, Garretts, Robeys, Fosters and others—built threshing machinery as well. They were mostly willing to allow credit to purchasers of good standing on very similar lines to the system which we now know as 'Hire Purchase'. We are apt to think of the Victorians as backward and old-fashioned, but they were mostly very shrewd, hard-working and progressive-minded people. Britain led the world in the business sphere, and was by far the greatest exporting country. And because so many steam inventors were British, those who built engines were supplying foreign markets as well as meeting the demand at home.

6

In the last chapter we saw how the final traction engine design for farming took shape. The decade 1860–1870 was also that in which steam ploughing went ahead. Earlier inventions had been tried and all but one were faulty. Some inventors tried to break away from the conventional method of ploughing—of turning over the soil in long narrow strips or furrows. Thomas Rickett, for example, saw no reason why rotating blades, power-driven behind the engine, should not be the perfect answer. The blades could be raised or lowered for deep or shallow cultivation, and at 90 lbs p.s.i. pressure could cultivate a seven-foot width of ground at a time. But the old problem of weight remained. At about eight tons the traction was poor on damp soil, and the wheels were liable to sink in, apart from their harmful consolidating effect. In addition, it was found when the engine was put through various tests that the cost per acre was more than by horse-drawn plough.

Rickett's invention was by no means the only failure. A Mr Savoury, to mention just one more, invented an engine where the winding drum revolved round the boiler barrel itself. It had several interesting features, including a separate donkey engine for pumping water into the boiler, and enclosed cylinders, motion and gearing. But it was cumbersome as well as ugly and, though designed to work as a pair, it did not take on. Other

6. This invention was known as the Walking Digger steam engine

inventors too brought out ploughing and cultivating machines which could not stand up to the quite severe tests which the Royal Agricultural Society organized. Such tests met a vital need for both the inventors and the farmers, who needed guidance when so many inventions were appearing. John Fowler's system of an engine and movable anchor at the other side of the field had so far proved the most economic. But it still wasn't good enough to give steam cultivation the real boost it needed.

Then Fowler hit on the idea of using two engines as a pair, one on either side of the field. He also patented a special winding drum, or winch, known as the 'clip drum'. This was for the wire rope, to which the two-way plough was attached, to be hauled back and forth across the field by each engine in turn. Wire ropes were, and still are to some extent, rather tricky and often dangerous unless made to keep taut and to wind on and off the drum evenly, without overlapping or crisscrossing. The clip was a sure guide for the rope, and by using a pair of engines, with the drum suspended beneath the boiler barrel, several problems were solved at once. To transfer the power of the crankshaft on top of the boiler to the winch fitted below it, a vertical shaft was fitted with bevel pinions at both ends. It was the lower end which engaged the teeth of the sprocket ring of the winch drum.

John Fowler exhibited his double engine system in 1864 at the Newcastle Show. The engines were not specially large and powerful, but they stood up to the tests, drawing a four-furrow plough so well that they made their inventor the leading exponent in the country. But a few months later, when still under forty, he died after a hunting accident. He had, however, laid a firm foundation for further development. His firm at Leeds improved still more on his inventions and orders began to come in from all over the world.

For the next sixty years Fowlers retained the supremacy they had achieved in those early days.

Ploughing sets or 'tackles', as they were called, were almost invariably used by contractors. We have seen how threshing sets, contractor-owned, went from farm to farm. Because ploughing sets were even more costly, it would be a very large-scale farmer indeed who could justify purchasing an outfit purely for his own use. There were of course instances of fairly prosperous farmers who could afford one mainly for their own work, and make it pay fully by ploughing some of their neighbours' fields as well. But contractors sprang up in all the corn-growing districts of Britain. They charged piece-work rates for ploughing, cultivating or mole draining at so much per acre, with the farmer supplying coal and water.

The land which responded best to steam working was the heavy or clay areas of eastern England. In due course there were contractors who owned a dozen or so sets, which would travel long distances, with crews living on the job. They too, were paid on an acreage basis, and of all the work in connection with steam, theirs was the toughest and most exacting. This was because it was seasonal work. Farmers were not keen to have them during the depths of winter, when the land was sticky and wet. There were five men to a crew, though one of them usually a lad. His job was to run errands, cook meals in the living van and generally relieve the drivers during what was often a sixteen-hour day. Two of the men were rough riders·on the plough or cultivator and the other two were engine drivers. Team work was vital and a system of blasts on the engine whistle became a code by which needs were met, and operations kept running. The really busy times were from harvest till winter clamped down, and a briefer spell in spring. Another task for

7. The driver's view of a Fowler ploughing engine

8. Fowler ploughing engine: a side view of the gears driving the winch drum which carried up to $\frac{1}{2}$ mile of wire rope

which ploughing engines came to be used was to dredge ponds and lakes, using a scoop between the two engines. I have also seen them used to clear woods and scrub land. The power on the end of their ropes is tremendous. It is enough to pull out sizeable trees by the roots, and having pulled out half an acre or so of small trees and bushes, the rope can be the means of encircling them and drawing them on to a huge fire.

9. The complete view of a Fowler ploughing engine in operation, showing the wire cable being paid out

Fowlers went on later to the building of general purpose traction engines and haulage locomotives, both road and rail. But they remained specialists in ploughing engines and achieved perfection in design and efficiency until, in the 1930s, they gave way to diesel power. Their 'steamers', however, were widely exported and some sent to grain-producing countries were specially designed as straw-burners. Russia was one of the principal importers, but in the panic food shortage of the 1914–18 war, large orders were given for engines to increase food production in England as well as for export to Russia, which was an ally. Some of the engines sent to Russia were lost at sea, and those that arrived were not paid for because of the Russian Revolution in 1917. The pair of ploughing engines I possess were built in 1917, part of a batch that was intended for Russia. They were the largest of the three classes of 18 n.h.p. (this rating is explained on p. 36), the others being 16 and 12 n.h.p. respectively. One of my 22-ton engines had stood in an open scrapyard for seven years, rusting away. And yet, once the pistons were freed, and a steam test was given, the quality of workmanship was so good that, apart from repainting, no repairs

10. Ploughs, mole drainers and cultivators: implements used by ploughing engines

11. An Aveling road roller at work about 1912

at all were needed, and after another ten years the engines run as smoothly as ever. Of all traction engines, built simply for service, without ornamentation, Fowlers' ploughers are in my opinion the most imposing and majestic.

Avelings of Rochester also achieved renown as specialists. They too began to meet the need for rollers and, though competition became fierce later on, they were the leading builders of road rollers by 1870. One of the first errors Thomas Aveling made was on the score of weight. No doubt he imagined that, up to a point, the greater the weight the better the consolidation of a road. At that time 'macadam' was the only durable surface.

It was of special granite stone, with a little soil, and when rolled one of the flat surfaces of each stone came face up to give a level, if dusty, finish. Avelings' roller of 1867 was of thirty tons' weight, and proved far too heavy. It consisted of a boiler and works on a huge frame between two large rolls of seven feet diameter. It worked in Liverpool for a spell, but further orders received (including two for New York) were for lighter machines. The advance in design which led to its full popularity produced a machine with smooth broad hind wheels and a wide single roll in front, which came out in 1871. This method, which enabled the hind wheels each to roll a strip outside that compressed by the front roll, was copied by nearly all other builders in time. They rarely went back to the 'tandem' type, but as late as 1924 the firm of Robey built a comparatively light model to meet the increasing need where tarmac was being used. They were curious little machines, chain driven, and, though not many were built, I was able to acquire one in good running condition in 1965. Rollers were the last steam engines to be worked on the roads of this country, and most of these remnants were Avelings. They carried the name 'Invicta' on the front saddle, with the crest of a rampant horse.

Steam rollers played a vital part in the Second World War. Wherever runways for aircraft were laid, they were used for both preparation and repair. In much-bombed Malta, they were specially well protected and cared for, because of the need to make runways and roads serviceable again. For a brief spell just after the war, new rollers were much in demand for export. Marshalls of Gainsborough were given an order for a thousand to be exported to India, and Avelings were still in demand until about 1950. Even now, one hears occasional reports of steam rollers still being used commercially from places as far apart as Spain and India.

A very few rollers are still in use in England, though perhaps now mainly for sentimental reasons. But there were some which held proud records for reliability and long service. Sixty years of useful life was quite possible where drivers cared for them, as many did. This caring, a feeling for engines as if they were domestic animals rather than man-made machines, still exists among drivers and owners, and is the main reason why so many steam engines have been preserved.

12. An Aveling roller at a Rally

7

It should not be thought that the age of experiment ended with the 1870s. Farm engines and rollers were by then reaching their final stages of development, and were being used to an ever increasing extent. But there was one aspect of steam power as to which inventors and builders were feeling very frustrated. This was in the use of steam for road transport, and so it had been for over forty years.

Earlier in this book I mentioned the pioneers, those who built steam carriages and wagons which failed to take on. Bad roads were only partly to blame; prejudice and other opposing interests were even more responsible for barring those early and quite promising steamers from the roads. Stage-coach owners disliked the coming of railways, but they fought even more against steam on the roads, and in this the new railway companies then became their allies. There was also much alarm about the frightening effect road engines would have on horses, as well as about noise and smoke.

So it was that in 1831 came the first of many Acts of Parliament and by-laws laid down by local authorities which penalized and hampered the new source of power. In every respect those who used steam for any kind of road haulage were heavily penalized. As one example, if coal was steam-hauled, the toll or levy for using the roads was 4 shillings per ton, against $3\frac{1}{2}$d per ton for haulage by horses and carts. It was an offence to travel faster than walking pace, and a man with a red flag had to walk in front of the machine, even if it was a farm engine on the move from place to place.

These regulations, which in some cases were quite stupid, applied only to England. Most foreign countries were much more realistic, but because most engines had their origin in England, makers were obliged to seek special permission when testing on the roads, before offering them for export abroad. The foreign market was a wide open door. As the reputation of portable and farm engines was already established, foreigners were keen to use steam for road work as well, including passenger-carrying vehicles.

Accounts of various tests of the 1860–80 period make very interesting reading. Some machines were driven over long distances, but most intriguing were the variations made in design. Two engines built by Nathaniel Grew could certainly not have been fully tested in England because they were designed for use on ice in Russia. The first, built in 1860, had spring skids at the rear, with cylinders beside the footplate and a connecting rod to the driving wheel in the centre. In front was a single skid or runner to which the steerage handle was fixed, but one wonders how such a method could be effective on slippery ice. Be that as it may, Mr Grew sold the machine to a customer in Moscow and by 1862 sent out another which worked on the frozen river Neva in the north of Russia. This had larger skids in the front and none at the rear, but the single driving wheels were impelled by front outside cylinders, rather like a railway engine. Both engines weighed twelve tons!

Tuxfords built a curious road engine in 1862. This had seats over the broad rear wheel, and an undertype engine, with both driver and steersman placed above the two front wheels. In the same year Taplins of Lincoln exhibited an engine with a large driving wheel in an almost central position. This was driven by a pinion direct on the crankshaft end to a sprocket or spur wheel. The double cylinders were enclosed in the smoke-box, and both driver and steersman stood on the footplate. They were back to back, since the engine was designed to run backwards. It was reckoned to haul loads of fifty tons or more. Robeys, also of Lincoln, designed what they called 'The Highway Locomotive', of less cumbersome appearance, but with very large sprocketed wheels, with the driver dwarfed behind them and the steersman perched high up in front of the chimney.

Vertical boilers were occasionally used in the designs of that period. In 1867 R. W. Thomson adopted one for a machine designed for use overseas, on which rubber-tyred wheels were fitted. The idea of rubber bars for gripping was not so good in practice as in theory, for rubber by itself did not last long. For several years, with Thomson as designer and various established firms as builders to his patents, many road steamers were produced. They were compact and manageable and would have met with much greater success but for his failure to use rubber wheels in a trouble-free way. As it was, both Robeys and Burrells exported quite a number, mainly to the West Indies, to work in the plantations. The use of Thomson engines at work in India, commercially, led to the British Army in occupation there becoming interested in them for military purposes. R. E. Crompton, who founded the big electrical firm, was instrumental in this and with Thomson designed engines for hauling military 'trains' by road.

In turn this led to another engine being designed and built by Robeys for use at Woolwich Arsenal. It was a larger affair, still with the vertical boiler, and was named 'Advance'. When tried out in 1870 it hauled two passenger-carrying vehicles, with forty-five people on board, at over six miles per hour, taking a one in nine hill at four to five miles per hour.

13. 'Ravee': a pioneer road locomotive built for work in India

The boiler was of Thomson's own design; it was known as the 'pot boiler', and had a copper vessel fixed inside for increased heating surface. But insufficient grate area made steaming unreliable when the engine was working hard. So Ransomes built another, named 'Chenab', a three-wheeler destined for use in hauling passengers in India, but even the modified pot boiler was not a success.

At this period the demand for machines for export was so keen that at least three firms—Robeys, Ransomes and Burrells—went all out to produce a really reliable road haulage engine. A Burrells model, back to front three-wheelers with rubber tyres, went to Russia.

Having forsaken the pot vertical boiler, they went in for the horizontal locomotive type instead and both the Greek and Turkish governments placed orders. Crompton also discarded Thomson's boiler but used Fields' type, another vertical, for 'Ravee', to be built by Ransomes for India. When completed it was given a real test by travelling from Ipswich to Edinburgh and back. At one point it touched twenty-five miles per hour, but many troubles were encountered. The return journey of 424 miles took just over sixty-one hours running time—an average of just under seven miles per hour—but the total time, including stops, was nine days. By modern standards, such results would be ludicrous. But it was described at the time as an epic journey and it encouraged Crompton to return to India to make the necessary arrangements to receive and operate a fleet of similar vehicles. After many modifications his rubber-tyred wheels had after all proved a success. What was learned about rubber and its qualities paved the way for its universal use on the roads later on. Until then it was an almost untried material in this respect. Crompton's endeavours led to a wider use of solid rubber tyres and though Dunlop was credited a few years later with the invention of pneumatic tyred wheels, the first patent was taken out as early as 1845.

Rubber tyres had several advantages over iron, especially for military purposes. A military term, 'Sapper' (the nickname for the Royal Engineers), was applied to an engine built by Avelings for the Army in 1871. It was light, about five tons, and was priced at only £300. It could be used for hauling military supplies and field guns and to drive pumps, but scarcely ever for moving the troops themselves. They were expected to march. The appearance of this engine belied its period, for it was not unlike engines built fifty years later. Thanks to the Road Locomotive Society, an 1871 Aveling, which had these

14. A small general purpose traction engine built by Aveling and Porter

improvements but with iron wheels, was restored and presented to the Science Museum in 1954.

Avelings' works were close to Chatham Naval Dockyard, which may have helped in obtaining orders for crane engines. They could be used wherever portable cranes were helpful, and could also be used for hauling wagons. Cranes were not difficult to fit to an engine, but it took ingenuity to apply steam power to their lifting capacity. The jib was most often fitted to the front of the engine, for though a few were at the rear, the drive and control for it were awkward. These crane engines were, however, basically closer to the standard farm or general purpose engine.

The road locomotive had its uses and its limitations. It was found that light railways were more reliable and economic, and after about 1880, thousands of miles of narrow gauge lines of from $1\frac{1}{2}$ ft to 3 ft 3 in were laid in the less accessible parts of the Empire as well as in foreign lands.

The export of road locomotives for abroad did not however come to a stop because of the spread of light railways. In the 1880s the firm of McLaren of Leeds built a series of fairly fast engines. These were four-wheelers, with an enclosed cab. One type was for hauling a passenger bus or car in India at a speed of eight miles per hour, the other for more general haulage in France. This was one of a very successful batch of three with a working pressure as high as 175 lbs p.s.i., and it was reckoned that they each travelled 15,000 miles a year on a daily goods service between Lyons and Grenoble. Most builders by this time had forsaken three-wheelers and adopted the worm gear for steering, which became standard on traction engines and rollers. This made footplate steering possible simply by turning a handle which pulled in the chains fixed to the front axle according to

15. This Fowler, with its patent valve gear, was the forerunner of a larger and faster engine for road haulage

16. A representative of the lesser known makes of traction engine, this Allchin was built in 1911

17. A road haulage locomotive built by Robeys of Lincoln

the required turn.

    Fowlers were always in the forefront with new designs, and having more or less perfected their ploughing engines, they turned their attention to road haulage as well. They also built a military engine, and called it the 'Artillery Siege Train' crane engine. The crane was over the front wheels to make it more manoeuvrable; it had a short wheel base and a vertical boiler. Before this oddity, however, they brought out a haulage engine which had the startling feature of rear wheels twelve feet in diameter. These did not carry

18. A haulage engine with a rear-mounted revolving crane

19. These 12-foot driving wheels on a Fowler engine were a means of combating soft road conditions

the weight of the engine, for their axles were far too high. The idea was that a low slung axle carried the weight, which it transferred by a toothed disc on to a rack inside the great outer wheel. This was the drive, and the big wheel certainly gave it more grip on the road, but it proved far too cumbersome as a machine, and no more were made. A smaller one, with nine-foot wheels, was supplied to an estate in Norfolk. Though Fowlers had faith in large wheels for good traction, it was their seven-footer that became more popular for road work.

The firm of Fowell, at St Ives, Huntingdonshire, were not in a large way of business, but they too came out with some novel ideas. As a boy, living not far away, I used to stand and watch men at work on engines in their yard until I was turned away. That was nearly sixty years ago, when such firms were still flourishing. What I did not see was a very unusual engine they built in 1877, which was driven by a stout coupling rod direct on to the rear wheel from a cranked disc. This was the patent of a Mr Box, but though several of these engines were built, none have survived as relics.

Both the Leeds firms of Fowlers and McLarens were building heavy haulage locomotives by 1880, and not only for export. In the industrial north the demand was growing for engines to haul large and weighty equipment which could not be handled by rail, with its restrictive width. The famous 'Great Paul' bell was such a load. It weighed seventeen tons, and was hauled by a steam engine from Loughborough to St Paul's Cathedral by a Fowler engine. McLarens also built a few engines with four wheel drive. The first had the now conventional overtype cylinders and motion, but the second was an all geared drive undertype. The idea was good, but both failed because they were too heavy and ungainly.

With double cylinders and higher pressures, the road locomotives of the 1880s were also sprung, and were fitted with injectors as well as pumps. The needs of industry were such that permission to use them could not be withheld, though restrictions on speed, especially, were applied. For some contracts, such as for ships' boilers and other large indivisible loads which could not be taken by rail, special wagons had to be built. These helped to spread the weight, but routes had to be carefully selected to avoid the risk of bridges collapsing or the fouling of railway viaducts, to say nothing of narrow streets and corners too sharp to negotiate with extra long loads.

20. An 1875 Robey hauls a new boiler to be installed in a Suffolk brewery

By the 1890s the point was reached where the road authorities were forced to allow steam on the roads. The growing needs of industry were largely responsible for the relaxation of the many stupid and short-sighted restrictions which had accumulated over the previous sixty years. The coming of the motor car added weight to the outcry for more freedom. It is well known that the first internal combustion vehicles came under the 'red-flag law', which demanded that they should keep to walking pace behind a man holding a flag. But it is not often appreciated that it was the makers of steam engines who were the pioneers in many respects. Nor is it realized that less than forty years after freedom came at last in 1896, fresh laws against the steam engine put an end to its use on the roads—laws which, it is said, were influenced by the new motor industry's hostility to steam.

8

A little more has to be added to these brief historical notes. Before going on to give more details of how traction engines work, and how a great deal of enjoyment can be and is being gained from those in preservation, it is necessary to know how final designs for specific purposes emerged. The greater freedom on the roads following the 1896 Act of Parliament naturally gave a great fillip to both builders and users of steam engines. Farm engines and rollers were not greatly affected by the change in the law. They had no need to increase speeds, since it was well known that for road-rolling very low speeds were more effective. Both types were generally without springs and had only two gears. Low gear was under 2 m.p.h. and in top, 5 m.p.h. was the usual limit.

For road haulage this was not good enough. Top gear would often bring a speed of 10 m.p.h., and this made springing a necessity. Road locomotives were of much heavier construction, and carried belly tanks for extra water. These tanks were slung under the boiler barrel, but a valve allowed it to run into the other tank beneath the footplate and coal tender because it was from this that water was drawn to replenish the boiler. Rubber tyres were a costly extra at first, but later on, when the price of rubber fell, they became more or less standard fitting.

These road locomotives for heavy haulage weighed from fourteen to eighteen tons.

21. A Showman's Road Locomotive, with dynamo, built by Garretts of Leiston, Suffolk

Because of the increasing variety of tasks they were able to perform, they were sometimes made to customers' requirements. They pulled trailer loads of heavy industrial equipment, stone blocks for building, roadstone, timber, girders, grain and many other items. Travelling showmen also became interested. The first to realize the value of a smart engine for hauling his equipment from fair to fair, was Jacob Studt from South Wales. Extra brasswork, a decorative canopy and crested side plates round the motion were

22. An ornate and well cared for Burrell Showman hauling its train of fairground amusements

included in a Burrell he bought in 1889. From this time Burrells made a speciality of Showman's Engines, and competition in splendour between different showmen up and down the country proved to be very good business for the famous Thetford firm. My father used to stay as a boy with relatives living within a stone's throw of their works, and saw many a new engine emerge during the 1880s.

Jacob Studt was also the first showman to use electric lighting. The Army had set the pattern, for dynamos fitted to an engine could be used to provide search or arc lights. They were first placed over the motion, but they suffered from oil splash, so they were moved to the front of the chimney. This position added to the length of the engine and made it necessary to have a complete canopied cover. These dynamos were of course driven by a belt from the engine flywheel. A showman in a big way might have more than one engine, but when they were on the road, the most ornate and specially equipped one would head the procession. Restrictions were imposed later on as to how many vehicles were allowed on tow, but till then it was a very splendid sight to see up to nine vehicles in tow behind a magnificent engine with its twisted brass and gleaming paintwork. It is small wonder that 'Showmans' in preservation are now the most highly prized and cared for traction engines.

Although an unadorned engine, and often horses too, were used to bring up the less important fair equipment and living vans, there were some much smaller road locomotives with the showman finish of canopy and brasswork. These too were kept in splendid condition, and a few specimens by such makers as McLaren, Garrett, Ransome and Burrell still survive. Far more of these 'tractors', as they were called, were used as replacements for horses in the heyday of steam—1890–1930. They were usually of about five tons in weight, but Taskers' 'Little Giant' was only half that weight and Fosters' 'Lincoln Imp' was little more. They had a good turn of speed and could haul loads up to twenty tons in weight, depending somewhat on road surface and levels. They were iron-tyred (as were their load vehicles) until nearer the end of their era, but they were prodigious workers, and travelled long distances very economically when furniture etc. had to be moved.

All these engines, including rollers and general purpose farm tractions, were fitted with rope drums holding a wire cable. It was coiled in a grooved inner wheel or 'drum' behind a driving wheel through which two heavy pins were able to keep the drum turning with the wheel. If these pins were taken out, only the drum could revolve when the engine was in gear. In other words the rope drum, or winch, could only be used when the engine was stationary. This was in fact where it came in most usefully, and the pull it could exert was enormous. A farm engine, for example, could haul its equipment through a muddy yard or pull quite large trees out by the roots. As a boy I saw an old house being levelled to the ground with a wire rope from an engine, and have myself raised a steel water tank on a thirty-foot tower from the horizontal on the ground up to its required vertical position. Timber hauliers used rope drums extensively, as did showmen and other users of the road. And it was always a safeguard should the engine itself run into trouble in soft or boggy ground.

In spite of the upsurge of steam on the roads during the 1890s, inventors were still active. The potential of steam power was by no means exhausted and improvements in detail were still coming into use. Necessity being the mother of invention, it is easy to imagine how the notion of a self-propelled truck or wagon arose, now that the 'horseless carriage' had come with the motor car. A traction engine for road haulage was of course a locomotive carrying out the same function as a railway engine—that of drawing other vehicles. They were serving a vital need, but more than one inventive engineer was thinking how handy it would be if a vehicle could be fitted with a small steam engine and still leave room for a pay load.

Steam waggons are not strictly traction engines, but no matter. They come into the

23. Two Foden timber tractors in a trial of strength at a Rally

general category sufficiently to be of real interest as one of the later developments, and several are preserved. A spate of steam waggons came out in the early 1900s. Such famous names as Thornycroft and Leyland were amongst the pioneers who very soon forsook steam in favour of petrol. Several of the well-known steam engine builders persisted longer, including Foden, Clayton, Ransome, Garrett, Burrell and Robey, but of these only Fodens of Sandbach met with real success, as far as widespread sales were concerned. At any rate, several Foden waggons and tractors with the same short horizontal boiler remain in preservation. They continued building them until 1931, and I was lucky enough to come by one of that year, with solid tyres. By the 1920s they had become very efficient, with Ackermann instead of chain steering, and could cruise comfortably with a five ton load at 16–18 m.p.h. Even so, the engine occupied almost half the total length of the vehicle, and it weighed five tons empty.

One other make of steam waggon survived longer, and production did not cease until about 1950. It was the Sentinel, which, with its vertical high pressure boiler, could carry heavy loads. The last of this famous line, the Super Sentinel, could reach 55 m.p.h., and but for the stumpy chimney, with its smoke, it could scarcely be recognized as a steamer. It made a faint whispering noise, and it was very modern in appearance. But the increasing efficiency of diesel engines, with their lower running costs, put an end to further development of steam on the roads. Sentinels were much more of a specialist job than other firms who made a variety of engines. This specializing allowed far more economy in manufacture, but the firm did produce one other type in 1929—a speedy and very powerful timber tractor. It had two engines, one for winching and the other for propulsion, but at that time of depression it cost £2500 and only twelve were built. Only two were preserved and one of these is in the Bressingham Museum. The demand for traction engines and road haulage locomotives fell off rapidly during the later 1920s, and by the mid 1930s, production came to an end.

With reasonable care, a traction engine had a working life of at least forty years. As late as 1940 many were still in regular use, despite the increasing popularity of internal combustion and diesel engines. Wartime needs checked the change-over, but it was sad to see these old-timers being cut up for scrap once the emergency was past. Owners wishing to change over to oil were persuaded by ranging scrap dealers to part with their old machines and because thousands of engines had become unwanted, prices offered were low. A ten-ton traction engine often brought no more than £10 as scrap. They were often cut up on the spot, or towed away to a breaker's yard, and during the period 1946–1950 it is estimated that ninety per cent of all road-using engines disappeared in this way.

I bought my first engine in 1947 and paid £50 for it because it had been overhauled, with a new set of tubes. The function of boiler tubes is to increase the heating surface by drawing heat and smoke from the firebox through the boiler barrel to the smoke-box and chimney. They are of drawn steel and those for traction engines vary in diameter from $1\frac{1}{2}$ inches to $2\frac{1}{2}$ inches and in number from about thirty to sixty. They fit into holes at each end, in what is called the tube plate, and become tight by means of special expanders which squeeze the thin tube outwards. They are subject to fire on the inside and boiling water outside when in use and have a limited life, because when they are not in use, rust eats into the metal. Hard water is also harmful, and an engineman would expect to have to renew tubes every two to five years. Soot had always to be kept down by using a special brush on a rod, and this was often a daily task.

Apart from tubes, moving parts such as cylinders, bearings, sprockets and axles need to be constantly supplied with oil to avoid wearing. Cylinders require a much heavier grade of oil, and it is forced in by a special mechanical lubricator, working off the stroke. The most expensive renewal happens when the inner shell of the firebox becomes thin; a replacement means complete dismantling. They can, however, be patched, especially with modern welding techniques, if one section only has become dangerously thin. It is the firebox, much more than the boiler barrel, that calls for the closest examination when the necessary annual insurance inspection is carried out.

Another important maintenance task is that of washing out the boiler. Most engines carry a maker's plate stating that a cleanse-out should be done every fourteen days, or every hundred working hours. Shale, mud and other water impurities collect with the conversion of water into steam. This impedes the flow of water in the narrow spaces between the firebox and outer shell of the boiler, if it is allowed to accumulate. Plugs and 'mud hole doors' are fitted for this purpose. It is common practice nowadays to add a chemical to the water which retards corrosion of the tubes and plating. Costs of repairs and replacements have risen enormously since steam engines became collectors' pieces. A new firebox, for example, costs as much today as a complete new traction engine would have done fifty years ago.

Most of the fourteen road-using engines in my collection were purchased at well under their original price when new. A general purpose traction engine cost in the region of £500 new at the beginning of the century. Many were sold for scrap in the 1940s regardless of condition for as little as £15 but by 1960 scarcity put up the price of those remaining engines which were capable of being restored and preserved, to around £150. By 1970 restored engines were fetching more like £1500, and since then prices have risen so that a Showman's in good order will change hands for £5000 or more. Even steam rollers, many of which were still in use in the 1950s, and thereby escaped the scrap man, have recently been sold for over £1000. Because the craft of engine building has died out, with patterns and other equipment gone, it would probably cost £10,000 or more to make a new engine.

This gives an idea not only of the value but of the appeal these relics of the steam age possess. I was fortunate in acquiring a collection when I did, and, adding railway engines of various types and sizes, I have been able to set up a Museum Trust. This ensures that the collection will not be dispersed, but will be on show to the interested public. On open days* several of the engines come to life and because the appeal of steam in action is so great, the museum can be self-supporting. Proof of the nostalgia for steam is given by the number of people who attend Traction Engine Rallies. These too have greatly increased. The first few were held in the 1950s, but now there are scores every year up and down the country. Some of the larger ones may include other attractions, such as a fair, models, demonstrations and relics other than steam. These extend over two or three days, and attract fifty thousand or more visitors, and it is not uncommon for ten thousand people to come to a one-day event. At these Rallies, one may usually see representatives of all types of engines. Steam rollers and farm or general purpose engines of various makes may predominate but a wagon or two, road locomotives and Showman's all contribute to what is an impressive sight. Such Rallies often begin with an opening parade, with perhaps thirty engines, all in steam, proceeding majestically round the field. Paint and brass gleam brightly despite the smoke and if anyone enjoys the sight more than the visitors it is the drivers of the engines, who are proud to be in charge of a live old-timer once more. Some may well have lavished days or weeks in preparation for the event, which gives such a splendid excuse to raise steam once more. They may go in for some of the competitions and perhaps win a prize or trophy. They may also be out of pocket. If

* The Museum and Gardens at Bressingham are usually open on Sundays from early in May to early in October, and on Summer Bank Holidays, from 1.30 to 6.30 p.m., and on Thursdays from late May to mid-September from 1.30 to 5.30 p.m. For further information apply to The Bressingham Steam Museum, Diss, Norfolk.

24. A grand parade at a Traction Engine Rally

the engine has to be brought to the Rally by low loader, because the distance was too great for it to travel under its own power, such costs are not always met in full by the Rally promoters.

Some engine owners, however, love to travel long distances under steam, even at ten miles per hour or less. Even the John O'Groats–Land's End run has been accomplished in recent years, though of course it took several days. It is gruelling work to keep a chain-steered engine on course, and the 3-speeder which can make ten to twelve miles per hour needs a steersman to supplement the driver's skill in keeping up steam and water and using it to best advantage. But such is the fascination of steam now that the age of steam has passed, that owners and drivers revel in the work which engines involve. It is not gruelling to them, for it is a labour of love, just as it was to restore the engines and bring them back to life.

Acknowledgements

No. 2 and No. 3 are taken from the manufacturers' original copper plate blocks, now in the author's possession. Acknowledgements for the remainder are made to the following: The collection of the late Mr Gerald Dixon: Nos. 1, 4, 6, 7, 8, 14, 17, 20, 24; *The Times*: No. 5; John Tarlton: No. 9; Richard Burn: Nos. 10, 12, 22; The Cromptonian Association: No. 13; Road Locomotive Society: Nos. 11, 18; John Fowler & Co., Leeds Ltd.: Nos. 15, 19; *Free Press*, Bury St. Edmunds: No. 16; *Maldon and Burnham Standard*: No. 21; J. L. Griffin: No. 23.

Index

'Advance', 53
Allchin, 59
'Artillery Siege Train', 59
Aveling, Thomas, 30–2, 48
Avelings, 55, 57
Avelings' Road Roller, 48–9, 50
Avelings' 'Sapper', 55
Aveling and Porter, 56

Box, Mr, 61
Boydell, James, 26, 27
Boydell Patent Wheel, 26–7, 28, 30
Bray, William, 27–8, 30
Burrells, 26, 27, 40, 53, 54–5, 68
Burrell 1909 model, 9, 10
Burrell Showman, 65–6, 71

'Chenab', 54
Claytons, 25, 26, 40, 68
Clayton and Shuttleworth, 31
Crompton, R. E., 53, 55

Dixon, Gerald, 9
Dunlop, 55

'Farmers' Engine', 22–4
Foden, 68–9
Fosters, 40
Fosters' 'Lincoln Imp', 66
Fowell, 60–1
Fowler, John, 25–6, 43–4
Fowlers, 58, 59–60, 61
Fowlers' Ploughing Engine, 44–7
Fowlers' 'Artillery Siege Train', 59

Garretts, 26, 40, 64, 66, 68
Garrett Portable, 24–5
Grew, Nathaniel, 52

'Highway Locomotive', 53
Hornsbys, 17, 25

'Invicta', 49

Leyland, 68
'Lincoln Imp', 66
'Little Giant', 66

McLarens, 57, 61, 66
Marshalls, 40, 49

Ransomes, 14, 40, 54, 66, 68
Ransomes' 'Chenab', 54
Ransomes' 'Farmers' Engine', 22–4
Ransomes' 'Portable', 13, 14, 25
Ransomes' 'Ravee', 54, 55
'Ravee', 54, 55
Rickett, Thomas, 41
Robeys, 40, 49, 53, 54, 59, 68
Robeys' 'Advance', 53
Robeys' 1875 model, 62
Robeys' 'Highway Locomotive', 53

'Sapper', 55

77

Savoury, Mr, 41
Sentinel, 69
Showman's Road Locomotive, 64
'Steam Elephant', 28–9
Studt, Jacob, 64–5
Super Sentinel, 69

Taplins, 52–3
Taskers' 'Little Giant', 66
Taylor, James, 28–9
Thomson, R. W., 53, 54, 55
Thornycroft, 68
Trevithick, Richard, 12
Tuxfords, 26, 52
Tuxford 1858 model, 26

Walking Digger, 42
Watt, James, 8, 11